HERESY AND SCIENCE

IN THE MIDDLE AGES

Prof. Camillo Di Cicco, M.D.
American Association for the History of Medicine.
European Academy of Dermatology and Venereology.

Giordano Bruno

Between the January 1589 and the spring of 1590, at Helmstedt, Giordano Bruno wrote *"The Lulliana Medicine"*. This work consists of a practical application of the Lullian System in the medical astrology.

The text, which incorporated large sections of *"Explanatio compendiosaque applicatio Artis Raymondi Lulli"* (1235-1315) by Lavinetha, opens with the premise that health is determined by the balance of four elements, fire hot and dry, the air hot and humid, the water cold and wet and the earth dry and warm.

Regarding the astrological studies applied to the medical art, Bruno gave to the world a vision of nature alive full of magic and spirituality. Since the astral forces are in harmony with human nature, Bruno wanted to find in them a therapy possibility, developing a real own astral medicine, taking care of the bodies of the men directly through their minds.

Giordano Bruno must be regarded as the staunch defender of his own ideas, as inviolable dignity of wisdom.

"Brave, Fendi space with my wings and fame does not make me bump worlds drawn from false principles, under which we would be locked in a prison imaginary as if everything was surrounded by walls of iron"
(De Immense)1591.

In the early morning hours of Friday, February 17, 1600, one of those processions which were all too familiar to Rome was seen wending its way to the Campo dei Fiori, the place where the Holy Mother Church burned her heretical sons. Giordano Bruno was led to the pile, clad as a heretic, *"...his tongue was imprisoned because of his wicked words".*

Bruno was bound to the stake and the hungry flames began to lick at his flesh. But not one sigh of agony escaped from that noble breast. When, at the last moment of his torment, a crucifix was held before him, he turned his eyes away. When the fire had died out his ashes were dumped into the Tiber river.

Bruno's works were placed on the *"Index Librorum Prohibitorum"* (1603).

INDEX LIBRORVM
PROHIBITORVM,

CVM REGVLIS CONFECTIS
per Patres a Tridentina Synodo delectos,
auctoritate Sanctiss. D. N. Pij IIII,
Pont. Max. comprobatus.

VENETIIS, M. D. LXIIII.

"I fought a lot; I thought I could win, but fate and nature repressed my study and my efforts. But it is already something to be on the battlefield because to win depends very much on fortune. But I did as much as I could and I do not think anyone of the future generation will deny it. I was not afraid of death, I never gave in to anyone, I chose courageous death instead of a coward's life".

G. Bruno. De Monade (1591)

Giordano Bruno works:

Animadversiones circa lampadem lullianam
Ars memoriae
Ars reminiscendi, Triginta sigilli et triginta sigillorum explicatio, Sigillus Sigillorum
Articuli centum et sexaginta adversus huius tempestatis mathematicos atque philosophos
Artificium perorandi
Cabala del cavallo pegaseo
Camoracensis Acrotismus seu rationes articulorum physicorum adversus peripateticos
Candelaio
Cantus Circaeus
Cena de le Ceneri
Centum et viginti articuli de natura et mundo adversus peripateticos
De compendiosa architectura et complemento artis Lullii
De gli eroici furori
De imaginum, signorum et idearum compositione
De innumerabilibus, immenso et infigurabili
De l'infinito, universo e mondi
De la causa, principio et uno
De lampade combinatoria lulliana
De magia
De magia mathematica
De monade, numero et figura

De progressu et lampade venatoria logicorum
De rerum principiis et elementis et causis
De triplici minimo et mensura
De umbris idearum
De vinculis in genere
Figuratio Aristotelici physici auditus
Idiota triumphans - De somnii interpretatione
Lampas triginta statuarum
Libri physicorum Aristotelis explanati
Medicina lulliana
Mordentius - De Mordentii circino
Oratio valedictoria
Oratio consolatoria
Spaccio
Summa terminorum metaphysicorum
Theses de magia

Buno's works were placed on the Index Librorum Prohibitorum (1603).

Marguerite of Hainault, la Porete.

The first of June 1310, in Paris, the heart of medieval culture, to the mill the Saint-Antoine, after the burning of 54 Templars sentenced for heresy, Margherite la Porete was burnt alive as heretic together with his book *"The miroir des simples âmes"* (the mirror of simple souls), of which the Church ordered the destruction.

"Le miroir des simples ames anienties et qui seulement demeurent en vouloir et desir d'amour" today represents one of the vertices of religious thought speculative, a manifesto of the nobility of the soul.

Marguerite of Hainault, called "la Porete", was a beguine, or member of a female urban religious order that practiced charity and good works in medieval cities. Marguerite, a woman of the leisure class and of great culture, knew the Scriptures, the theory of humors of Galen, was born in the region of Hainaut in Belgium, then part of the Holy Roman Empire, and between 1296 and 1306 she wrote his book translated into four languages.

It was initially attributed to St. Margaret of Hungary (1242-1270), but in 1306, Marguerite was named as the true author.

Accused of heresy for its content by the Inquisition, the text was burned in his presence. Sentenced to silence of his ideas, she was later found "Relapsa" (relapse) in 1308, and again accused of heresy.

Tried in Paris by the Bishop of Cambrai, Philip de Marigny, was sentenced and 1 June 1310 was burned at the stake, on the Place de Greve. Witnesses report the time of his courage and his dignity in front of the gallows.

The Book of Marguerite, whose full title was *"Le Miroir des simples ames here anienties et seulement demeurent en Vouli et d'amour desire"* is the most ancient mystical text in vernacular French.
\]
The literary genre is that of the mirror, very popular during the Middle Ages.

*"You who would read this book, If you indeed wish to grasp it, Think about what you say, For it is very difficult to comprehend; Humility, who is keeper of the treasury of Knowledge And the mother of other Virtues, Must overtake you. Theologians and other clerks, You will not have the intellect for it, No matter how brilliant your abilities,
If you do not proceed humbly" (Mirror).*

The *"Mirror"* says the growth of the spirit in its desire for union with God through seven stages in the dialogue between Love, Reason and Soul.

*"Humble, then, your wisdom Which is based on reason, And place your fidelity In those things which are given By Love, illuminated through Faith. And thus you will understand this book
Which makes the Soul live by love. p.79". Mirror.*

The danger of the ideas of Marguerite, for the Inquisition, is apparent particularly in some passages of his work. The notion that the soul undone by love of the Creator can give nature all you want, coincides with the precepts of the Brethren of the Free Spirit.

These, in the sense of being filled with the Holy Spirit, believed to be perfect, and therefore in a position to commit any action without having to incur the sin, following the dictates of St. Paul:
"Everything is pure to the pure" (Letter to Titus 1:15).
Moreover, in his letter to the Galatians, St. Paul wrote:
"Quod si Spiritu ducimini, non estis sub lege",
"But if you are led by the Spirit, you are not under the law"
(Gal., 15, 18).

Spiritual Franciscans, like the Beguines and Begardi, were sourly persecuted by Pope John XXII who commissioned their repression at the dreaded inquisitor Bernard Gui in the first half of 1300. Therefore in *"Practice officii inquisitionis heretice pravitatis"*, Bernard Gui (Royères, 1261 - Laurou, 1331), inquisitor at Toulouse from 1307 to 1323, describes in detail the techniques that should be used by an inquisitor during the interrogation of a heretical. The text, addressed solely to the Inquisitors, it is certainly crucial in defining the "modus operandi" of the Inquisitor and is divided into five parts: 1) thirty-eight formulas related to the citation and arrest of heretics. 2) fifty-six acts of grace. 3) forty-seven judgments. 4) the powers and prerogatives of the inquisitors. 5) interrogation technique.

Is reported, by the *"Practice officii inquisitionis heretice pravitatis"* by Bernardo Gui, snippets of questioning of a heretic by an Inquisitor (translation from the Latin):

[Inquisitor] When a heretic is first brought up for consideration, he takes a confident attitude, as if it was secure in his innocence. I ask him why he was brought to me.
[Heretical]. (He says, smiling and polite) "Lord, I'd be happy to know the reason from you".
[Inquisitor]: You are accused of being a heretic, and that you believe and teach differently from the creed of the Holy Church.
[Heretical] (Looking up at the sky, with the air of the utmost faith) Lord, Thou knowest that I am innocent of this, and I've never held a faith different from that of true Christianity.

11th April 1309:
"Condemnation of the Book of Marguerite Porete"
Notarised by Eveno Phil and Jacob of Virtue. This document reports that twenty-one theologians were convoked by William in his capacity as Inquisitor and asked to advise on several articles extracted from "a certain book" (*The Mirror*). No mention is made of prior condemnations. After some deliberation the articles are declared heretical and erroneous. Porete's name is not given, nor is the title of her book.

The originality of the "Mirror" is to devise a new form of self-consciousness, inconceivable for a woman in this dark period, which inevitably will lead to individual freedom, characteristic of modern culture.
\|
Marguerite proclaim with force and not give up his ideas, thereby sacrificing his own life at stake June 1, 1310.

St. Bernard of Clairvaux (1090 – August 20, 1153) wrote:
"Faith must be persuaded and not imposed".
"Fides suadenda non imponenda"

Marguerite la Porete\]

All the documents relating to Porete's condemnation (including the *Nangis* and *Frachet* chronicles, the *Grand Chronicles of France* and *Ly Meur des Histoirs*) are held in the Archives Nationales, Paris, layette J428.

In 1986 they were published with an exegesis by Paul Verdeyen in *"Le Procès d"inquisition contre Marguerite Porete et Guiard de Cressonessart (1309-1310)"*.

Henry Charles Lea also translated many of the documents in his *"A History of the Inquisition of the Middle Ages"*, III vols. (London: Sampson / Rivington, 1887, 1988), Vol. II, pp.575-78.

Lea's offerings were reprinted in 1889 by Paul Frédéricq in *"Corpus documentorum inquisitionis hareticae pravitatis Neerlandicae"*
(Gand, 1889) Vol I, pp. 155-60.

The following passages have been translated from the Latin by Richard Barton from Henry Charles Lea, *A History of the Inquisition in the Middle Age*, 3 vols. (NY: Macmillan, 1922), 2:575-578; and from *Corpus documentorum inquisitionis haereticae pravitatis Neerlandicae*, ed. Paul Fredericq, Hoogeschool van Ghent, Werken van den pratischen leergang van vaderlandsche geschiedenis, 5 (Ghent: J. Vuylsteke, 1896), 156-160. \]

Richard Barton from Henry Charles Lea, *A History of the Inquisition in the Middle Age*, 3 vols. (NY: Macmillan, 1922), 2:575-578

The University of Paris Provides Consultation in the Case of Marguerite In the name of Christ, Amen. Let it be made known to everyone through the present public instrument [ie., public document], that we, the below-written notaries, had gathered together at St-Mathurin de Paris at the request and command of that religious man, Brother William of Paris, inquisitor into heretical depravity in the kingdom of France by apostolic authority; we were namely, venerable and discrete men Simon the dean, Thomas of Belliaco,

William Alexander and John of Ghent, all canons of Paris; Peter of St Denis, Gerard of St Victor, James abbot of Carolocus, Gerard the Carmelite, John of Poilly, Laurence the prior of Vallis Scolarium, Alexander Henry the German, and Gregory of Lucca, all of the Order of Hermits of St Augustine; John of Mont Saint-Eloi, Ralph de Hoitot, and Berengar, all of the Dominican Order; John of Claromarisco, Nicholas of Lyra, and James of Esquillo, all Franciscans; James the Cistercian and Roger of Roseto, masters of Theology. Having gathered us all together in the same place, the same inquisitor asked advice from the masters concerning what was to be done about a certain book which he displayed there and from which many articles had been extracted, put on display, and demonstrated to the masters, in the manner that follows.

The first of these articles was this: "That the annihilated soul gives licence to the virtues and is no more in their servitude, since it does not have use for them, but [rather] the virtues are subject to the will." Item, the fifteenth article was this: "That such a soul [ie., one annihilated in love of God] does not care about the consolations of God nor does it care for His gifts; and it neither ought to care for nor is able to care for [such things], because it is wholly intent upon God; and thus its intention towards God can be impeded." Having first held deliberation with the other masters listed above, Simon the Dean of Paris, in the name of and with the will, assent and concord of all the above said masters, made a response to the request for consultation that had been directed to them. He stated that it was and is their advice that such a book, in which the said articles are contained, ought to be exterminated as heretical, erroneous, and contemptuous of heresies and errors. Done in the aforementioned place, in the year of the Lord 1309, on the 11th day of April. [this date actually gives the year 1310, April 11].

\|
Corpus documentorum inquisitionis haereticae pravitatis Neerlandicae, ed. Paul Fredericq, Hoogeschool van Ghent, Werken van den pratischen leergang van vaderlandsche geschiedenis, 5 (Ghent: J. Vuylsteke, 1896), 156-160.

The Inquisitor Consults Canon Lawyers on the Case of Marguerite la Porete, May 30 1310. To all those who will inspect the present letters, William called Brother, archdeacon of Laudonie in the church of Saint-Andrew in Scocia, Hugh de Bisuncio canon of Paris, John de Tollenz canon of Saint-Quentin in the Vermandois, Henry de Bitunia canon of Furne, and Peter de Vaux curate of Saint-Germain-l'Auxerrois of Paris, all ruling Paris according to canon Law, give greetings in the name of the author of salvation. You ought to know that the venerable, devout and discrete man, Brother William of Paris, of the Order of

Preachers, who was deputed by the authority of the Apostolic See to be inquisitor into heretical depravity within the kingdom of France, communicated to us [the facts of] the case which follows, and consulted with us about it as is described below. The case truly was such: when Marguerite called Porete, who was suspected of heresy, remained disobedient and rebellious, not wanting to respond to questions nor swear an oath before the inquisitor concerning those matters which pertain to the office of inquisitor (which office had been granted legally to William), the same inquisitor nevertheless held an inquiry concerning her and, through the deposition of many witnesses, found that the said Marguerite had composed a certain book containing heresies and errors.

Moreover, this book had been solemnly and publicly condemned and burned as such [ie., as heresy and error] by the order of the reverend father, lord Guy, the late bishop of Cambrai; in his pronouncement Bishop Guy had also ordained that if she should again attempt such things as were contained in the book either in writing or in speech, he condemned her and relinquished her to the secular justice for judgment. The same inquisitor [William] also found that she had confessed, first before the inquisitor of Lotharingia and then before the reverend father lord Philip, then bishop of Cambrai, that [even] after her initial condemnation she had retained the said book and other books; he also found that the said Marguerite had sent the said book (containing in its similitude the same errors) after its condemnation to the reverend father lord John, bishop of Chalons by the grace of God. He also found out that she had sent the book not only to the said bishop, but also to many other simple people, to Beghards, and to others of similar status. The consultation resulting from these matters that the aforesaid inquisitor made with us was done in this way: namely, he asked us whether on account of such things the aforesaid Beguine ought to be judged to have relapsed? [this was an important distinction; anyone could fall by mistake or ignorance into heresy, but if, once warned, you returned to that heresy (ie., relapsed), then the authorities could be sure of your evil intentions].

We, moreover, responding canonically to the aforesaid consultation as lovers of the Catholic faith and as those professing the truth, decided that the same Beguine [Marguerite] was condemned by the truth of the facts listed above, and that she thus ought to be judged a relapse and as such ought to be handed over to the secular court. As testimony for this matter, we affixed our seal to the present document. Done in the year of the Lord 1310 on the Sunday after the feast of St John [ie., on May 30 1310] before the Latin Gate.

In 1570, in a letter of Pope Pius V to Philip II of Spain on how to deal with heretics, reads:

"Reconciliation never: never pity; exterminated those who submits himself and those who resist endless; persecuted to the death, kill, burn, all go to fire and blood, as long as it avenged the Lord, much more than its enemies, are your enemies".

Hildegard Von Bingen (1098-1179)

Hildegard von Bingen (1098-1179) was born at Bermersheim, Magonza (Meinz), was a remarkable woman. At a time when few women wrote, Hildegard, known as "Sybil of the Rhine", produced major works of theology, medicine and visionary writings. Hildegard composed music and spoke of Christ as God's song, *"Symphonia harmoniae caelestium revelationum"*.

When few women were accorded respect, she was consulted by and advised bishops, popes, and kings. In medicine she used the curative powers of natural objects for healing and wrote treatises about natural history and medicinal uses of plants, animals, trees and stones.

Her scientific books contain more than 2,000 remedies and health suggestions.

In the work "*Liber simplicis medicinae*" printed in 1533 and called Physica, she tells of the basic qualities, the medicinal value and the proper application of 230 plants, 63 trees, 45 animals.

In "*Liber compositae medicinae*", called Causae et curae, Hildegard speaks of the external world, but always with reference to human health (the kinds of water that are safe to drink); on illnesses and their causes, on cures, and finally on symptoms to be looked for.

Hildegard von Bingen is the first composer whose biography is known, founded a vibrant convent, where her musical plays were performed.

Hildegard's music heard is frequently an antiphon:
An antiphon (Greek ἀντίφωνον, ἀντί "opposite" + φωνή "voice") is a response, usually sung in Gregorian chant, to a psalm or some other part of a religious service, such as at Vespers or at a Mass.

Hildegard

1 ANTIPHONA - O coruscans lux
O coruscans lux stellarum,
o splendidissima specialis forma regalium nuptiarum,
o fulgens gemma,
tu es ornata in alta persona,
quae non habet maculatam rugam.

Tu es etiam socia angelorum
et civis sanctorum.

Fuge, fuge speluncam antiqui perditoris,
et veni in palatium Regis.

PSALMUS 1 Beatus vir
Beatus vir, qui non abiit in consilio impiorum, et in via peccatorum non stetit,
et in chatedra pestilentiae non sedit.

Sed in lege Domini voluntas ejus,
et in lege ejus meditabitur die ac nocte.
Et erit tamquam lignum, quod plantatum est secus decursus aquarum,
quod fructum suum dabit in tempore suo:
Et folium ejus non defluet:
et omnia quaecumque faciet prosperabuntur.

Quoniam novit Dominus viam justorum,
et iter impiorum peribit.
Gloria Patri, et Filio, et Spiritui Sancto.
Sicut erat in principio, et nunc, et semper, et in saecula saeculorum. Amen.

The Hildegard's musical language of the liturgy was that of the Gregorian chant.

The music of Hildegard pervaded and animated of spiritual strength, should certainly be considered an expression of the late Gregorian chant,

The essence of the music of Hildegard coincides with her spirituality.
Hildegard shows an intense contemplative experience that leads to a deep bond, authentically mystic, with God.

Hildegard dies the September 17, 1179, she has been beatified and is frequently referred to as St. Hildegard.

Revival of interest in this extraordinary woman of the middle ages was initiated by musicologists and historians of science and religion

Miniatur aus dem so genannten Lucca-Codex des "Liber divinorum operum": Hildegard am Schreibpult, um 1220/1230, Biblioteca Statale in Lucca

HISTORY OF MEDICINE

Originating as divine and supernatural, Greek medicine changed and moved toward analysis and logical thinking during the period 800 B.C. to 460 C.E.

Thales (636-546 B.C.), philosopher and scientist, undertakes examination about the laws of nature and physics. He supposed that water (moisture) was the first element from which the world was formed.

Empedocles (Agrigento c.495- c.435 B.C.) philosopher and physician, who lived in Sicily, wrote *"On Nature"* and *"On Purification"*. Its system was based on the interaction of the four elements (fire, air, earth and water), called by him "rhizomata" (roots) under the influence of love and hate (attraction and repulsion). He studied circulation of the blood and atmospheric pressure, foreshadowed view of evolution. He was the founder of Italian medicine.

Empedocles (Agrigento c.495- c.435 B.C.)
Philosopher and physician, was the founder of Italian medicine.

In 460 B.C. was born Hippocrates on the island of Kos. He wrote the famous *"Hippocratic Oath"*. Its sixty works are collected in the Corpus Hippocraticum. Hippocrates supported the humoral theory in which our body is governed from four various humors (blood, yellow bile, black bile, flegma), which, when arranged in different ways, lead to health or the disease.
Medicine owes to him the art of clinical inspection and observation, therefore he may justly called the "Father of Clinical Medicine". In Regimen in Health, Hippocrates wrote:

"A wise man should consider that health is the greatest of human blessings, and learn how by his own thought to derive benefit from his illnesses."

The Ancient Greek world comprised Greece, Turkey, Egypt, and parts of Italy and consequently these concepts were widened in every part of the Mediterranean Area.
Abu Ali al-Hussein Ibn Sina famous with the name " Avicenna", was born in Persia in the 980 to Afshana near Bukhara, Uzbekistan. By the age of eighteen years he was possessor of an immenses philosophical-scientific understanding and undertook the medical profession.
Avicenna was studious of Hippocrates and Galen therefore developed the theory of four humors and the derived

complexions. Avicenna was also known to fuse philosophy and medicine. Like Aristotle and Plato in reasoning, his infuence on western medicine was enormous, primarily through a work that soon became the predominant medical texts in universities: "The Qanun fit at-tibb."

Translated from Gerardo of Cremona in Latin with the name of *"Canon Medicinae"*, the work, which systematized ancient medical thought, is composed of five books.

The first book describes theoretical medicine, the second simple medicine, the third has diseases and their localization, the fourth has general diseases, and the fifth has pharmacology.

The Qanun results be connected undoubtedly to the Aristotelian tradition. The Canon, in the words of Dr. William Osler, has remained "a medical Bible for a longer time than any other work" and is said to have influenced Leonardo da Vinci.

Canon Medicinae

Avicenna wrote also a Poem of medicine *(al-Urguza fi at-tibb)*, a medical treaty in verses where the medicine is defined like the art to conserve the health and eventually to recover the disease appeared in the body.

Avicenna was the first doctor to detect the presence of sugar in diabetic urine, but last but not least, he was also poet, philosopher in addition to physician.

Avicenna died in June 1037 at fifty-eighth years, his body was buried in Hamadan, Iran.

MEDIEVAL MEDICINE

Medieval medicine gave great importance to the planets as influences in disease. The influence of the stars began at birth and influenced complexion. The continuous flow of celestial forces could change the course of disease.

The position of the planets was important in choosing the moment in which to begin cures or carry out a bloodletting. In "*Regimen against the Plague*", Siegmund Albich (1347-1427) invites readers not to think about the plague because that was sure to cause its appearance.

During the Middle Ages, the prevailing popular attitude in medicine was dictated by S. Ambrogio (334-378) who declared: *" the garnishments of the medicine are contrary to celestial science, contemplation and prayers".*

The epidemics assisted religious hysteria, including flagellating processions, in the deep conviction that to obtain healing only required calming divine anger. An erudite Italian physician, Petrus d'Abano, 1246 or 1250 - 1316 or 1320?, denied the existence of spirits and ascribed all miracles to natural causes. Brought before from the Inquisition in 1306 as a heretical, a magician, and an atheist, he defended himself and was absolved.

Petrus was accused a second time but, while the trial was preparing, he died. He was condemned "post mortem" and his body was disinterred and burned. A colleague hid the remains of his body after cremation and the Inquisition burned his effigy in the public square of Padua.

His important works were:
"Compilatio physonomiae, Lucidator dubitabilium astrologiae, De imaginibus and Conciliator differentiarum philosophrum et praecipue medicorum", this last excellent work was an attempt to reconcile Arab medicine and Greek natural philosophy.

Heresy and Medicine of the Templars.

At Paris, 18 March of 1314, on the island of the Seine in front of the Garden real, Jacques de Molay, the last Great Master of the Templars, and Geoffroy de Charny, preceptor of Normandy, were burned as heretics.

Thus finishes the history of the Knights of the Temple after two centuries. The Templars would have been in possession of the most hidden secrets of alchemy. They were first to use the Hypericum on burns and hurts from cut, like antiseptic, astringent, healing, and in order to improve humor of the soldiers that remained immobilized in bed for months.

The Templars created a mixture with pulp of Aloe, pulp of Hemp and wine of Palm called "Elisir of Gerusalem", with therapeutic and nourishing properties, they used the Arborescens Aloe for its antiseptic, bactericidal and fungicide actions and for its capacity to penetration in the deeper layers of the skin.
Robert Anton Wilson, in his book on the Templars, asserts that they used the hashish and practiced a shape of Arabic Tantrism; the doctrine of enlightenment as the realization of oneness of one's self and the visible world, combining elements of hinduism and paganism, including magical and mystical elements.

The authors of Holy Blood and Holy Grail, Baigent, Leigh and Lincoln, comment that the Templars need to treat wounds and illness, made them experts in the use of drugs and the Order in advance of their time regarded epilepsy not as demonic possession but as a controllable disease. Interestingly, cannabis is the safest natural or synthetic medication proven successful in the treatment of forms of epilepsy.

The esoteric inheritance and the alchemical-spagyrics acquaintances were handed from the Templars to the Crocifers. From these Orders, that one of Saint Giacomo or Jacobite managed many Hospitals during the XV century. To the Jacobite monks, in quality of experts in the cure of the diseases of the skin, the task was entrusted to cure the wounded soldiers during the Crusades, in the Hospitals of Malta and Cyprus. To them, in fact, was attributed the capability to create miraculous ointments.

In such historical context it must estimate the work of the Templars concluding with recognizing that they, anticipating the times, had a modern vision of the Medicine and, although were considered heretics and consigned to the fire, recently a document has been recovered in Archives Vaticans from the studious Barbara Frale that demonstrates as Pope Clemente V secretly pardoned Templars in 1314, acquitting their Great Master from the heresy accusation.

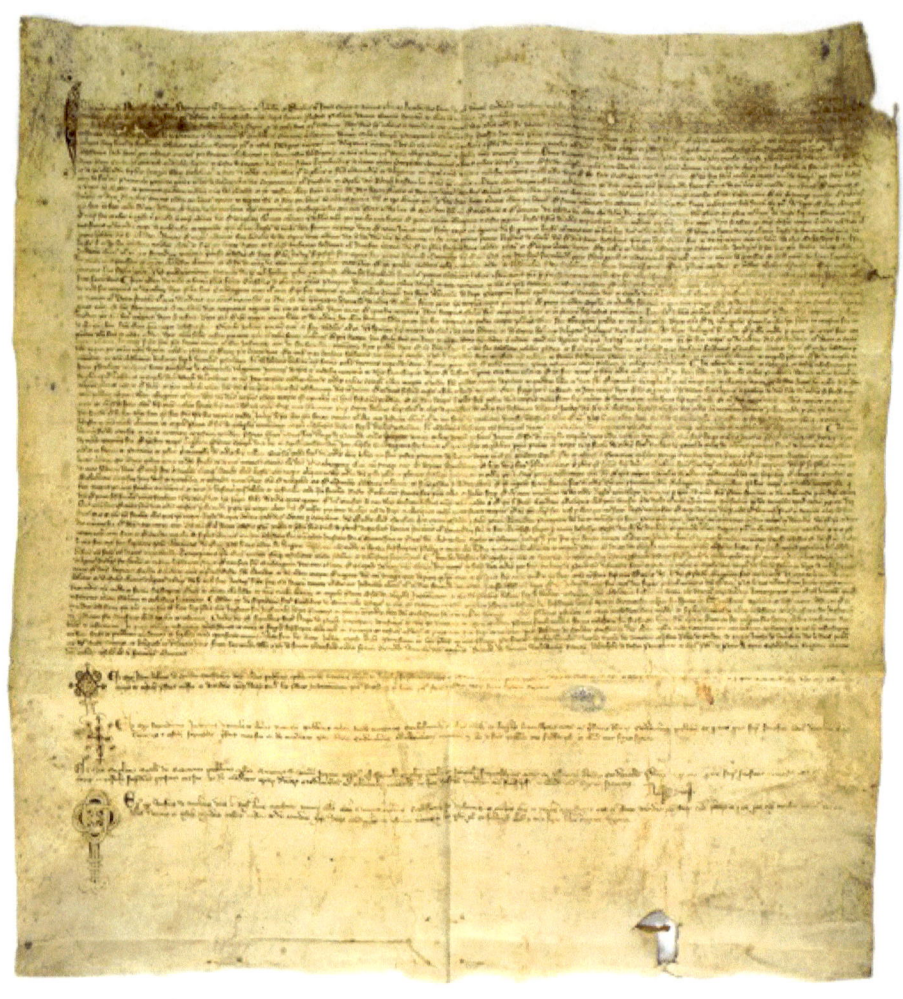

"Absolution's parchment for Templar Leaders including Jaques de Molay by the Apostolic Commission "ad inquirendum" of the 3 papal legates Bérenger Frédol, Etienne de Suisy and Landolfo Branacci
Chinon, Tours diocese, 1308 august 17-20. Archivum Secretum Apostolicum Vaticanum, Archivum Arcis, Armarium D 218".

Templars Burning Ancient depiction
(miniature from a French manuscript of the XV century)

Bible-Exodus 22:17 "..... do not to leave to live, those which practice magic". During the plague the witch hunts became frantic and inhuman. Many women were indicted for their independence, culture, particular attractiveness and suffered excruciating torture just for the fear of difference. The ability to be able to heal with simple ointments made of herbs, as was customary in the Middle Ages, it was used to point out them as sorceresses and witches. The Penal Code branch of Roman law, valid until 1200, operated under the adversarial system and this had prevented hitherto arbitrary trials, based only on accusations towards women named as witches. With the advent of the inquisitorial system in the subsequent period, it is totally delegated to the magistrate, protected by the church, the opportunity to indict even on the basis of their "beliefs, ideas and feelings".

The Inquisition also defines what is the corpus delicti, the minutes of the proceedings and the notice of investigation. With the subsequent introduction of torture in order to obtain the confession of the offender, the Court of the Inquisition acquires an uncontrollable power.

To the Inquisitors belonged to a "share pontifical" of confiscated property, and this favored the occurrence of abuse and unnecessary torture, in order to obtain a confession of the witches only for the confiscation of their assets and subsequently that of their families.

The historical and ecclesiastic Spaniard Juan Antonio Llorente (1756-1823) was one of the greatest historians of the Spanish Inquisition. Refers to the countless victims of Tomas de Torquemada (1420-1498), the Dominican inquisitor throughout the Spain and the colonies, that in the period of his term, which lasted 15 years, inflicted thousands of death sentences, life imprisonment and confiscation of assets.

In France, Nicolas Remy (1530-1616) historian and writer, author in 1582 of *"Demonolatry"* text concerning to witches and demons, he condemned 600 witches. Jean Bodin (1530-1569) philosopher, economist and jurist, wrote in 1580 *"The démonomanie des sorciers"* legal manual for the torture of witches. In Germany Benedickt Carpzov (1639-1699), accuser of witches, author of the book *"Practical Rerum Criminalorum",* explains the extreme torture that the inquisitor must be achieve during a process to obtain a confession. A Bamberg (Bamberg), city extracircondariale (Kreisfreie Stadt) of Bavaria in Germany, in the years 1626 to 1631 occurred in hundreds of casualties and was created in 1627 by Prince Bishop Johann Georg II Fuchs von Dornheim (1586-1633), the prison of the witches, "Drudenhausche". In this prison Johannes Junius, mayor of the city, was tortured and killed. In the letter that he wrote to his daughter before he died, we can understand how the process was carried out by the Inquisition. Below are some fragments of the letter written in German by Johannes Junius (translation in english):

I came to prison, I was tortured innocent, innocent I must die. For those who will come at home, nor do you need to become a witch or be tortured for so long for something that comes out of the their imagination, and, God have mercy, designes something to say. I want to tell you how things went for me ...

Dear child, six witnesses have testified against me at the same time: the chancellor, his son, Neudecker, Zaner, Hoffmaisters Hopffens Ursel and Else, all false, through the compulsion, everyone told me and pleaded (forgiveness) for the 'the love of God ......
When the executioner brought me back in jail, he said, 'Lord, please, for the love of God, confesses something, whether it is true or not. Invents something, because you can not bear the torture that will make you suffer, and even if you resist, a torture will follow another until you say to be witched.

In 1485 the Dominican inquisitor Bernard Rategno wrote a manual for the persecution of heretics *"Lucerne Inquisitorum Haereticae Pravitatis"*. The concept of demonic possession is thus codified in 1486 in germany by the Dominican friars J. Sprenger and HI Kramer in the *"Malleus Maleficarum"* (The Hammer of Witches). This work reached the thirty-five thousand copies printed and appears to be in this period, for relevance and diffusion, the second book after the Bible.

The text, translated from Latin into Italian, French and German, had 29 editions. In this work it is stated unequivocally, with the blessing of the church, the bond that is created between the devil and the woman. In this work is presented in a meticulous way the *"modus operandi"* of an inquisitor in the practice of witch-hunting. The text presents the techniques on how to extort a confession of a heretical, also states that the word female was derived from Latin, Fe Minus, that the woman is an inferior being.

"The pope and the Inquisitor"
Jeans Paul Laurens, 1882
Musèe des Beaux Arts (Bordeaux)

St. Paul in his first letter to Timothy 2,11-15, says: *"The woman learn in silence with all submissiveness. Don't allow a woman to teach, nor to usurp authority over the man, but to be in silence. Because first Adam was formed, then Eve, and Adam was not to be deceived, but the woman was deceived and became a transgressor, she can be saved in childbearing, if they continue in faith, the charity and holiness, with modesty".*

"The "Malleus Maleficarum" teaches the various methods of torture to be carried out during an interrogation, and finally the painstaking research on the body of marks, signs produced by the devil, in order to proceed with the condemnation and the consequently purification with the fire. *"The woman is an inferior being created by God not in His image and likeness. It corresponds to the natural order that women serve men"*. (St. Augustine, 354-430).

In this period was very dangerous for a doctor defend his scientific thesis and therefore to oppose itself to the dictates of the Inquisition. The Dutch physician, Johann Weyer, in Latin Wierus, 1515-1588, violently criticized the *"Malleus Maleficarum"* and the witch hunt with scientific arguments in *"De Praestigiis Daemonum et Incantatiponibus ac Venificiis"* (On the Illusions of the Demons and Spells and Poisons, 1563) and in the *"Liber De Lamiis"* (Book of Witches, 1577). In *"De Morbo Irae curatione et ejus et Philosophica Medica"* 1580, Weyer was the first physician to use the term "mentally ill" to describe a woman accused of practicing witchcraft.

He considered that the "maleficia" of the witches be solely due to medical factors, and therefore the confessions of witches, obtained after numerous tortures, were only a disorder linked to the uterus, which Weyer called "melancholy".

Weyer believed that witches were women only fragile, mentally ill, and although in this period of history the greatest inquisitors denigrated his works, he succeeded, with his scientific work, to give a great impetus to the study of mental illness since the '600 and be considered the father of forensic psychiatry.

It is impossible to quantify the number of processes of the Inquisition occurred in those years and their convictions: fire, life imprisonment, confiscation of property. Joseph II, Emperor of Austria, ordered that all documents of the Inquisition of Milan in the years 1314-1764 were burned, and behaved in the same way the officials of the Holy Office for documents relating to the processes between the years 1772-1810 in both Italy and Spain, also including the Colonies. In this way, the opening of the Vatican Archives by Pope Leo XIII, was unable to provide definitive data on the true extent of the dark period of the Inquisition.

The sources are the most varied, the Italian Encyclopedia refers to a million victims of this persecution in all European countries between 1575 and 1700, while in a recent interview, the theologian and priest Hans Kuhng has spoken of about 9 million witches killed from 1484 to 1782.

During the Enlightenment, in the new society that was created, the frantic witch hunt and the stake, after confessions obtained under atrocious torture, were considered barbarity and Voltaire writes in this regard: *"The witches have ceased to exist when we have ceased to burn them".*

Hieronymus Fracastorius (1478-1553)

Hieronymus Fracastorius

Girolamo Fracastoro (also known as Hieronymus Fracastorius), was born in 1478 in Verona, at that time still part of the Republic of Venice, to a noble family. He studied at Padua University, where he graduated in 1502. At the same University, he was assigned the chair of Logic and Philosophy.

His teacher was the doctor-philosopher Pietro Pomponazzi and his study colleagues were Andrea Navagero, who became a noted historian, along with Pietro Bembo and Gaspare Contarini, both of whom became Cardinals.

Medicine was his passion but he was also a humanist and a scientist, he was interested in astronomy, mathematics, physics, botany, geology, geography, and even composition of verses. He was contemporary and friend to Nicolaus Copernico (Copernicus).

To Pietro Apiano he affirmed that the comets' tails always appear along the direction of the sun, but in the opposite sense of it. He also described an instrument used in astronomy, realized in later years by Galileo Galilei: the spyglass.

As a doctor, he's considered one of the founding fathers of the modern medicine: he hypothesized that infections are caused by germs, with the ability to multiply inside the organism and to infect through the breath and different other forms of transmission.

Considering the acquired prestige in the field of the medicine, he was named, Archiatra Pontifical, and in the second time, the main Doctor of the Trent's council (1545), under the Pontificate of Paul III°.

Such charge, allowed him to be one of the craftsmen of the same council, wished from Holy See. The Church wanted to move the council from Trent to the city of Bologna, wich was nearer to Rome and outside from the Germanic Empire.

The occasion came when, in February of 1547, he diagnosed a worrisome epidemic of typhus fever bursted in Trent. The Papal legates, decided therefore, to transfer the Assizes to Bologna and the Pope confirmed it.

Works

In the 1521, Fracastoro wrote to Cardinal Pietro Bembo some letters, describing an unknown illness coming from the new world and cause of epidemics, that he named for the first: Syphilis.

In August of 1530 he published in Verona an epic poem in three volumes about it.

The work had 50 editions in Latin language and about 60 editions in other languages: German, French and English. In this poem is narrated the history of Sifilo, a young shepherd who offended Apollo, God of the sun, that punished him with a terrible illness that irremediably destroys the beauty.
Really the author composed two version of the work, both of them in Latin and in prose, dedicated to his friend and University study colleague: Cardinal Pietro Bembo.
One of the two essentially literary, with the title: *"Hieronymi Fracastorii Syphilis siue morbus gallicus"* easier understanding and wide diffusion; and the other in form of essay, entitled *"Hieronymi Fracastorii Syphilidis sive de morbo gallico"*, that results to be a scientific monograph, compiled for the Doctors; where the pathology is described in a detailed way, in its symptomatology, diagnosis and therapy

This last version has been found in 1939 by Francesco Pellegrini and preserved in the civic library in Verona. Fracastoro's work, was diffusedly known in Italy in its translation in vulgar Italian, "Syphilis sive de morbo gallico", published by Vincenzo Bernini, in the 1765:

"….. *Primieramente era mirabil cosa che l'introdotta infezione sovente segni non desse manifesti appieno se quattro corsi non compia la luna;*

che ricevuta nell'interno, tosto non appare al di fuor, ma si rimane per certo tempo ascosa, e poco a poco prende col pasto nutrimento e forza.

Da insolito torpor gravati e vinti da spontaneo languor, gli uomini intanto venian più tardi a l'opre e da pigrizia eran le membra tutte oppresse e vinte cadea dagli occhi il natural vigore e il natural color del mesto viso…..

sentiansi poi per grave duolo i membri, gli omeri lacerar e braccia e gambe, che la contagion, dappoi che corsa era di vena in vena e in un col sangue il nutritivo umore infetto avea discacciato al di fuor l'infetta parte da tutti i membri……."

> "……..*Tosto, pel corpo tutto, ulceri informi*
> *usciano e orribilmente il viso e il petto bruttavan, specie di*
> *malor*
> *novella……*". (p. 32-33).

"….. In the first place it was a marvelous thing that the introduced infection often manifest signs did not give the full four courses if it does not carry the moon;

that received in the interior, soon appears to outwardly, but remains for some time hidden, and gradually takes over meal nourishment and strength.

Burdened by unusual torpor and won by spontaneous languor, meanwhile the men came later in the deeds and laziness
the limbs were all oppressed and won, the natural force fell from his eyes and the natural color of the sad face …..

"……. Soon, for all the body, shapeless sores
came out and horribly, threw in the the face and chest,
a kind of new illness". (p. 32-33).

It has not verified if Fracastoro, Doctor and Poet, created this name, Sifilo, from nothing, or extrapolated it, from others already existing.

In fact, several etymologies have been proposed by luminaries authors (Es. *Sunphileo*, rather, coming from the love according to Falloppia).

But everything is, the origin of this word; it's certainly possible to affirm, that the neologism had a great resonance and replaced very soon, all the other names with which, the disease was identified.

The literary choice of the poem, being praised by the contemporary, thanks to the style and the rich imagination, has been for the author, the opportunity to describe in the scientific way the Syphilis and the possible remedies with which, he believed that it can be care: mercury and guaiac, called "ligno sancto", an originates wood of the Antilles, green, with black striations, very hard and considered the heaviest wood existing in the world.

Fracastoro sustained that the syphilis (considered a divine punishment because contracted for the easy customs) could be eliminated only through a deep perspiration.

For this reason he administered the mercury, which being toxic for the salivary glands, it produced a powerful secretion that allowed the recovery.

In similar way, the "guaiac" acted, it also needs to remember, that to the tisanes and the infusions, drawn by the guaiac, the medical science, attributed, in that historical period, powers of Panacea.

In 1535, Fracastoro wrote *"De causis criticorum dierum"*, and in 1538 *"Dies critici vel de dierum criticorum causis"* texts in which he analyzed from the physician-philosophical point of view, the critical days of the illness.

It 's known also, his work "*Homocentricum*", in which he proposed an alternative to the cosmological-tolemaico system, taking back the system of the homocentric spheres. In 1546 the most genial work of Fracastoro was published about the medicine, entitled "*De contagione et contagiosis morbis et curatione*".

The Italian Doctor conceived, the existence of contagion's vectors, that called "seminaria", during an epoch in which the microbes weren't known yet.

He had observed, in fact, that the illness was transmitted both for direct contact among people (contactu), that through objects, as it happens for example with the garments (fomites), both to distance as in the case of the smallpox and the plague.

In this work Gracastoro wrote:" it seems that three different types of contagion exist. The first infects only for direct contact. The second acts in the same way but it leaves besides tinders, and this contagion can spread out through these (tinders), as for example: the scabies, the tuberculosis, the malignant stains, and similar.

Speaking about tinders, I intend garments, wood objects and things of such sort, that although they aren't contaminated, they can preserve, however, the original germs of the contagion and infect through them.

For the third, a type of contagion exists, that transmits the illness for direct contact, through the tenders and also to distance, for example: pestilential fevers, the etisies, some types of ophthalmia , typhus and similar.

These different types of contagions, seem to obey to a determined law; rather, those that infect the far objects, do it, both for direct contact and through the tinders, those that are contagious through the tinderses are also it for direct contact, not all are contagious to distance".

Fracastoro therefore, realized the existence of microorganisms able to transmit the infections, proposing a scientific theory on the germs, 300 years before the empirical formulation by Louis Pasteur and Robert Koch.

Nevertheless its astronomic, cosmological and cosmogonic knowledges, brought him to integrate the theory of the epidemic contagion, with the presence of the influential power of the stars, in to progress of the same epidemics.

In 1546 the work was published *"De sympathia et antipathia rerum-Liber unus"*, a text of natural philosophy in which Fracastoro affirms that in nature everything is connected from a natural and universal strength: it deals of the "sympathia" of everything for the part, and of the part for everything.

This strength is considered by the author, not in spiritual sense, but in physical and natural sense, according to the laws of the atomistical theory. in fact he explains, that are the flows of the atoms to establish the relationships among all the things of the world.

Fracastoro sustains therefore, the attraction between the similar things and the repulsion among those different. Fracastoro refuses the explanation of the phenomenons through hidden causes, because he believes that the appeal to them, isn't an attitude that suitable to an authentic scientist.

He believes that it's always necessary, to examine deeply the facts, so it's possible to elaborate wide inductive generalizations.

Fracastoro, therefore, sustains that in all of our scientific investigations, it needs to follow to the description and the evaluation of the phenomenons.

He also wrote, three philosophical dialogues 1) *Naugerius sive de Poetica* - 2) *Turrius sive de Intellectione* - 3) *Fracastorius sive de Anima*.

These works treat the theme of the human nature, examined in its cognitive abilities and in its relationships with the cosmos and with a supernatural reality, in front of which, the speculative trial appears inadequate.

The first work, treats the poetic's theme, and it is the most meaningful of the last year of the author's life, because he strongly defends the autonomy of the art.

With an Aristotelian vision, even if, with many elements of platonic nature, the poetry is considered as universal representation, as the action through which the idea is realized in its visible beauty.

It's recognized besides, the freedom of the artist,to transpose the universal beauty, inside to the objects, in sensitive forms.
In the "*Turrius sive de Intellectione*", a dialogue to "gnoseologico"(gnosis γνοσις- logos λογος) character, the author faces the intellect's theme, considering the logic as the tool of natural knowledge. The tradition of the philosophers of end 300 and 400, which had elaborated deep analyses about the different forms of the sensitive experience, it's taken back and developed, in particular way, about the perspective and the optics.

The last work of the author is "*Fracastorius sive de Anima*", a dialogue of psychological matter.

In this text the thought of Fracastoro manifests it, with a series of questions that the author do to himself, at the end of his experience of life, as doctor and philosopher. The work results to be incomplete.

Fracastoro died, from a stroke, the 8th August of 1553 to Incaffi, actually Affi, place near Verona his native city, where he had predominantly lived and practiced his activity of Doctor.

BIBLIOGRAPHY

"The trial of the Templars in the Papal State and the Abruzzi" (Città del Vaticano, Biblioteca Apostolica Vaticana, 1992).
Aries, P. 1985. Aries, P. 1985. Images of Man and Death. Harvard University Press, 271p.
Chronicon Monasterii S. Salvatoris Venetiarum Francisci de Gratia (1141-1380), ed. A. M. Duse, Venezia 1766, pp. 69-70.
L. Green Chronicle into History. An Essay on the Interpretation of History in Florentine Fourteenth-Century Chronicles Cambridge 1972.
A. Coville Documents su les Flagellants «Histoire littéraire de la France» 37 (1937), pp 390-411.
Tononi AG. La Peste Dell' Anno 1348. Giornale Linguistico de Archeologia, Storia e Letteratura 1884;11:139–52.
Horrox R, editor. The Black Death. Manchester University Press; 1994. p. 14–26.
Hecker JFC. The epidemics of the Middle Ages. London: Sydenham Society; 1844.
R. Guarnieri, Prefazione storica, in M. Porete, Lo specchio delle anime semplici , traduzione di Giovanna Fozzer, prefazione storica di Romana Guarnieri, commento di Marco Vannini, Edizioni San Paolo 1994, p. 39.
Alfred D. Berger, "Marijuana," Medical World News, July 16, 1971, pp. 37-43; reprinted in Marijuana Medical Papers.
B. Guenée Storia e cultura storica nell'occidente medievale. Bologna 1991, pp. 255-61.

V. Rutenburg Popolo e movimenti popolari nell'Italia del '300 e '400, introd. di R. Manselli, Bologna 1974, p. 109.
N. Biraben Les hommes et la peste en France et dans les pays européens et méditerranéens, voll. 2, Paris - La Haye 1975-76 (Civilisations et Sociétés 35-36), 2, p. 69.
Umberto da Romans, De eruditione praedicatorum, II, XCII, in Malato, medico e medicina nel Medioevo di J.Agrimi-C.Crisciani, Torino 1980.
Miller, Tanya Stabler "What's in a name?: Clerical representations of Parisian beguines (1200-1328)", *Journal of Medieval History*, 33, (2007), pp.60-86.
Paul Verdeyen, "Le Procès d'"inquisition contre Marguerite Porete et Guiard de Cressonessart (1309-1310)", *Revue d'Histoire Ecclésiastique*, 81, (1986), pp.47-94.
Anonymous *The Cloud of Unknowing: a book of contemplation the which is called the cloud of unknowing, in which a soul is oned with God*, Underhill, Evelyn [ed.,], 2nd ed., (London: J.M.Watkins, 1922).
Edsman C.M. "Mysticism, Historical and Contemporary", in *Mysticism: Based on Papers Read at the Symposium on Mysticism Held at Abo on the 7th-9th September 1968*, (Stockholm: Hartman. S.S., & Edsman. C.M., Almqvist and Wiksell, 1970).
Elliott, Dyan *Proving woman: female spirituality and inquisitorial culture in the later middle ages* (Princeton: Princeton University Press, 2004).
"Dominae or Dominatae? Female mystcisim and the trauma of textuality", *Women, Marriage, and Family in medieval Christendom: essays in memory of Michael M.Sheehan,*

CSB, Rousseau, Constance; Rosenthal, Joel T [eds.], Studies in Medieval Culture XXXVII, (Michigan: Medieval institute Publications, 1998), pp. 47-77.

Foucault, Michel *Discipline and Punish: the birth of the prison*, Allan Sheridan [trans.], 2nd Vintage ed. (1997), (New York: Vintage Books, 1995)

Geybels, Hans *Vulgariter Beghinae: eight centuries of beguine history in the Low Countries*, (Turnhout, Belgium: Brepols, 2004).

Gies, Frances and Joseph, *Women in the Middle Ages*, (New York: Barnes and Noble, 1980).

Given, James B. *Inquisition and Medieval Society: power, discipline, and resistance in Languedoc*, (Ithaca, New York: Cornell University Press, 1997).

Gooday, Frances "Mechthild of Magedburg and Hadewijch of Antwerp: A Comparison", in *ONS Geestelijk Erf*, 48, (1974), pp.305-62.

Greenspan, Kate "Autohagiography and Medieval Women's Autobiography", *Gender and Text in the Later Middle Ages*, Chance, Jane [ed.], (Gainsville: University Press of Florida, 1996).

"Stripped for Contemplation" *Studia Mystica* XVI: 1 (1995), pp.72-81.

Greven, J *Die Angrange über Beginin: Ein Beitrag zur Geschichte des Volksfrommigkeit und des Ordenswesens im Hochmittelalter*, Münster: (Westphalia, 1912).

Rosenthal, Joel. T [ed.] *Medieval Women and the Sources of Medieval History*, (Athens and London: The University of Georgia Press, 1990).

Szarmach, Paul *An Introduction to the Medieval Mystics of Europe*, (Albany: State University of New York Press, 1984).
Taylor H. O. *The Medieval Mind: a history of the development of thought and emotion in the middles ages* (1911), 4th ed., (Cambridge, Mass.: Harvard University Press, 1966).
Thomson, David. "Deconstruction and Meaning in Medieval Mysticism", C*hristianity and Literature*, 40, No. 2, (Winter 1991).
Tobin, Frank. "Medieval Thought on Visions and its Resonance in Mechthild von Magdeburg's Flowing Light of the Godhead", *Vox Mystica: Essays on Medieval mysticism in honour of Professor Valerie M Lagorio*, Bartlett, Clark; Bestul, Thomas, H; Goebel, Janet; Pollard, William [eds.], (Cambridge: D.S.Brewer, 1995), pp.41-53. Turner, Denys *The Darkness of God: negativity in Christian mysticism*, (Cambridge: Cambridge University Press, 1995)
Fracastoro, Girolamo Hieronymi Fracastorii Syphilis siue morbus Gallicus. - Veronae, 1530 mense Augusto.\/
Fracastoro, Girolamo Hieronymi Fracastorii Syphilis, sive morbus Gallicus. - Romae : 1531, mense Septembri (Impressum Romae : apud Antonium Bladum Asulanum.\/
Fracastoro, Girolamo Hieronymi Fracastorii Veronensis. De sympathia et antipathia rerum liber vnus De contagione et contagiosis morbis et curatione libri III. - Venetiis : [eredi di Luca Antonio Giunta], 1546 (Venetijs : apud heredes Lucaeantonij Iuntae Florentini, 1546 mense Aprili).\/

Fracastoro, Girolamo Hieronymi Fracastori Homocentrica, eiusdem De causis criticorum dierum per ea quae in nobis sunt. - (Venetiis : apud heredes Lucae Antonii Juntae, 1538 mense Januario).

Fracastoro, Girolamo In fugam Caroli V imperatoris. Hieronymi Fracastorii carmen. - [1552?]

Fracastoro, Girolamo Hieronimi Fracastorii Veronensis. De temperatura vini sententia. Consalui Barredae Hispani, sententiam perpendens libellus. - Camerini : apud Antonium Gioiosum, 1553 (Camerini : apud Antonium Gioiosum, 1553).\|

Fracastoro, Girolamo Hieronymi Fracastorii Veronensis Opera omnia, in vnum proxime post illius mortem collecta ... Accesserunt Andreae Naugerii patricii Veneti, Orationes duae carminaque nonnulla, amicorum cura ob id nuper simul impressa, ut eorum scripta, qui arcta inter se uiuentes necessitudine coniuncti fuerunt, in hominum quoque manus post eorum mortem iuncta pariter peruenirent. - Venetiis : apud Iuntas, 1555 (Venetiis : apud haeredes Lucaeantonii Iuntae, 1555).

Fracastoro, Girolamo Hieronymi Fracastorii Veronensis Opera omnia, in vnum proxime post illius mortem collecta: quorum nomina sequens pagina plenius indicat. Accessit index locupletissimus.- Venetiis : apud Iuntas, 1574.\|

Fracastoro, Girolamo Hieronymi Fracastorii Veronensis Opera omnia quorum nomina sequens pagina plenius indicat. Accessit index locupletissimus. Ex tertia editione. - Venetiis : apud Iuntas, 1584 (Venetijs : apud Iuntas, 1574).\|

Lettere di diuersi autori eccellenti. Libro primo [quintodecimo]. Nel quale sono i tredici autori illustri, et il fiore di quante altre belle lettere si sono uedute fin qui. Con molte lettere del Bembo, del Nauagero, del Fracastoro, et d %60altri famosi autori non piu date in luce. - In Venetia : appresso Giordano Ziletti, all'insegna della Stella, 1556.\|
Fracastoro, Girolamo [1478-1553]. Hieronymi Fracastorii Syphilidis, sive morbi gallici lib. III. Ioseph lib. II. item Carminum lib. I. Rutilii Claudii Numatiani Galli u.c. itineraria. - Antuerpiae : Apud Martini Nutij Viduam, 1562 (CIVVR).
Fracastoro Girolamo "Syphilis sive de morbo gallico", libri tres ad Petrum Bembum, Patavii, J. Cominus, II Editio, 1739.\|
Fracastoro Girolamo: "Opera Omnia", secunda editio, Venetiis, apud Juntas, 1754.
Fracastoro Girolamo: "Della sifilide ovvero del morbo gallico", libri 3, volgarizzati da Vincenzo Benini Colognese, Della Volpe, Bologna, 1765
Garrison: "Fracastoro", Science, New York, 1910.\|
Gangolphe: "Le lesioni sifilitiche preistoriche", Mem. Acad. Scien. Lyon, 1912
Fracastoro, Girolamo [1478-1553]. Hieronymi Fracastorii Syphilidis : sive de morbo gallico = introduzione, versione e note di Francesco Pellegrini. - [Verona] : edizioni di Vita veronese, 1956 (CIVVR, SLVR).\|
Laita, Pierluigi.The stipend of Girolamo Fracastoro, Doctor of the Trent's Council, 1545-1547. Verona, 1971 (CIVVR) M].

INDEX

GIORDANO BRUNO

MARGUERITE OF HAINAULT

HILDEGARD VON BINGEN

HISTORY OF MEDICINE

MEDIEVAL MEDICINE

HIERONYMUS FRACASTORIUS

www.ingramcontent.com/pod-product-compliance
Lightning Source LLC
Chambersburg PA
CBHW041105180526
45172CB00001B/123